Nina Znidaric

Reproductive success of two different populations of Tropheus moorii

Nina Znidaric

Reproductive success of two different populations of Tropheus moorii

Paternity analysis of F1 – hybrid population of two different populations of Tropheus moorii

Natural Sciences Series

Impressum / Imprint
Bibliografische Information der Deutschen Nationalbibliothek: Die Deutsche Nationalbibliothek verzeichnet diese Publikation in der Deutschen Nationalbibliografie; detaillierte bibliografische Daten sind im Internet über http://dnb.d-nb.de abrufbar.
Alle in diesem Buch genannten Marken und Produktnamen unterliegen warenzeichen-, marken- oder patentrechtlichem Schutz bzw. sind Warenzeichen oder eingetragene Warenzeichen der jeweiligen Inhaber. Die Wiedergabe von Marken, Produktnamen, Gebrauchsnamen, Handelsnamen, Warenbezeichnungen u.s.w. in diesem Werk berechtigt auch ohne besondere Kennzeichnung nicht zu der Annahme, dass solche Namen im Sinne der Warenzeichen- und Markenschutzgesetzgebung als frei zu betrachten wären und daher von jedermann benutzt werden dürften.

Bibliographic information published by the Deutsche Nationalbibliothek: The Deutsche Nationalbibliothek lists this publication in the Deutsche Nationalbibliografie; detailed bibliographic data are available in the Internet at http://dnb.d-nb.de.
Any brand names and product names mentioned in this book are subject to trademark, brand or patent protection and are trademarks or registered trademarks of their respective holders. The use of brand names, product names, common names, trade names, product descriptions etc. even without a particular marking in this works is in no way to be construed to mean that such names may be regarded as unrestricted in respect of trademark and brand protection legislation and could thus be used by anyone.

Coverbild / Cover image: www.ingimage.com

Verlag / Publisher:
AV Akademikerverlag GmbH & Co. KG
Heinrich-Böcking-Str. 6-8, 66121 Saarbrücken, Deutschland / Germany
Email: info@akademikerverlag.de

Herstellung: siehe letzte Seite /
Printed at: see last page
ISBN: 978-3-639-46787-1

Acknowledgments

During my work for the following Master thesis I had a lot of help and support from my family, friends and colleagues. Here, I would like to thank to:

Univ. Prof. Mag. Dr. Christian Sturmbauer for a professional supervision, support and for enabling me an exciting experience of the fieldwork in Africa.

Univ.-Ass. Dr. Kristina Sefc, Dr. Stephan Koblmüller and Dr. Steven J. Weiss for all professional advices.

Oliver Bock for supervision in the Laboratory.

Mag. Lisbeth Postl, Dr. Michaela Kerschbaumer and Dr. Karin Mattersdorfer for help and support during my work.

Dr. Katrin Winkler for the best advice in the Laboratory to use the "Type-it Microsatellite PCR KIT".

Christine Börger for support and good cooperation during my studies, for friendship and fun and pleasant time that we spent together in Africa.

Mag. Martin Viertler for long discussions and help.

Claudia Radler, my dear friend for long hours of sitting with me in front of computer and motivation in my crisis of microsatellite analysis and for recovering coffee breaks.

To all laboratory co-workers for every advice and cooperation.

To my parents, Silva and Stanislav Žnidarič and my brother Matej, who supported me on every step of this long path and made this possible.

Table of contents

1. Abstract

Microsatellite markers are commonly used to determine family relatedness. We analyzed the reproductive success in a pond hybrid population of two different parental populations *Tropheus moorii* (*Tropheus* "Nakaku" females, $n = 75$; *Tropheus* "Mbita" males, $n = 25$; and F1 – offspring, $n = 133$), based on 6 microsatellite loci. Using the pedigree software CERVUS, 109 (82%) parent–offspring pairs were identified. Moreover, within the F1 generation six F2 individuals were discovered and assigned to their parents from the F1 population. The study shows a strong correlation between number of mates and number of offspring. Moreover it suggests the occurrence of sneaking males, or alternatively females mating with more than one male in one brood due to the artificially high density in the pond environment. Both behaviors have not been observed in the wild. Moreover, female preference is suggested to be crucial for males' reproductive success, as two males had a much higher progeny in comparison to the others.

Keywords: Cichlidae, microsatellites, reproductive success, parentage assignment

2. Introduction

Cichlid fishes of East African Great Lakes Victoria, Malawi and Tanganyika count hundreds of endemic species which evolved within the last thousands to a few million years. Their enormous species richness and diversity made them a prime model for the study of evolution (Fryer & Iles 1972; Meyer *et al.* 1990; Stiassny & Meyer 1999; Kornfield & Smith 2000; Verheyen *et al.* 2003; Kocher 2004; Salzburger & Meyer 2004; Sturmbauer, Husemann & Danley 2011). In Lakes Malawi and Victoria, species richness can be explained to a large extent by the action of sexual selection which caused diversification of both spatially segregated (e.g. Danley & Kocher 2001; Knight & Turner 2004; Pauers *et al.* 2004) and sympatric populations (e.g. Seehausen *et al.* 1998; Seehausen & van Alphen 1999; Maan *et al.* 2004; but see Arnegard & Kondrashov 2004; Coyne & Orr 2004), in addition to natural selection. However, the much greater morphological and behavioral diversity of Lake Tanganyika cichlid fish species, some of which are substrate breeders, points to the long-term action of natural selection, pushing eco-morphologically

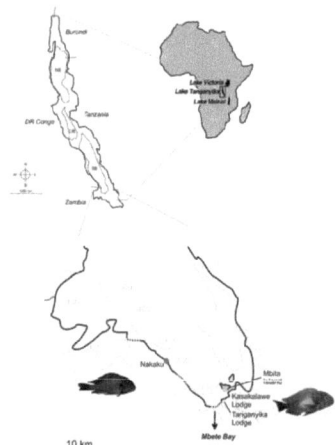

distinct species more and more toward extremes (Greenwood, 1984). In contrast to the younger lakes, the adaptive radiation is at highly advances stage in Lake Tanganyika, so that the pace of eco-morphological diversification has slowed down considerably and most new species evolve in allopatry as geographically separated sister species (Sturmbauer, 1998).

Fig 1. Map of Lake Tanganyika with sampling sites with Mbita (Egger et al. (2007)) and Nakaku individual (Caroline Hermann)

Lake Tanganyika (Fig. 1) lies in the central-East Africa and is with 9 - 12 million years

4

the oldest lake in Africa (Sturmbauer, 2000). With a surface area of 32.820 square kilometers (Coulter, 1991), it is the seventh-largest lake in the world and with the maximum depth of 1.470 kilometers, the world's second-deepest (Coulter, 1991). During history, all inhabitants of Tanganyika have been exposed to several lake level fluctuations promoting allopatric speciation. This was termed species - pump (Rossiter, 1995). Currently, Lake Tanganyika counts 200 valid cichlid species (Koblmüller *et al.* 2008a), all evolved from only few pioneer species which populated the lake at the beginning (Sturmbauer, 2000). Morphologically, also supported by molecular data (Koblmüller *et al.* 2008a), cichlids have been classified into 12 (Poll, 1986) or 16 (Takahashi, 2003) tribes.

The consequence of lake level fluctuations are many color morphs or sister species in several Cichlids, of which the genus *Tropheus* comprises about 120 (Schupke, 2003). Such a variety evolved genetically and morphologically due to geographical isolation (Sturmbauer, 2000).

Due to its large number of distinct populations, *Tropheus moorii* is one of the most interesting study species to target allopatric speciation. It feeds on epilithic algae and it is a strict rock-dweller with reduced ability of crossing open waters or sandy habitats. It is sexually monochromatic and both sexes inhabit and defend their territories against conspecifics and other individuals (Sefc, 2011). In the courtship period females normally form a temporary bond with one particular male (Yanigasawa & Nishida 1991; Sturmbauer & Dallinger 1995) and feed within the males' territory for weeks before spawning occurs (Sefc, 2011). Afterwards, females leave the male territory and mouthbrood their offspring elsewhere by themselves until their fry is released (Yanigasawa & Sato 1990; Sefc, 2011). Clutches consist of 2 – 14 eggs (Egger *et al.* 2006) and are fertilized by one male only, which is normally the females' selected partner (Sefc, 2011). Presumably, females are likely to choose the same partner repeatedly although no field data exist so far (Sefc, 2011).

Although sexual selection seems to be particularly important for the evolution of cichlid fishes in Lakes Malawi and Victoria, it is also likely to act in *T. moorii*, despite lack of dimorphism and polygamy (Sefc, 2011). The lack of color dichromatism was explained by the existence of additional social significance of color patterns at the population level, independent of the sex. Territorial observations found no correlation between body size and pairing success of males (Sefc, 2006) and laboratory observations detected no female preference for males' body size (Egger *et al.* 2010). However, mating success of males depends on a sizable territory (Yanagisawa and Nishida, 1991), therefore male - male competition has a great impact on male's reproduction which indicates on intrasexual selection. Sexual selection may, therefore still play a role in *Tropheus,* it could only be expressed otherwise (Sefc, 2011).

The present study addressed the reproductive success of *Tropheus moorii* in a hybrid breeding experiment carried out in large concrete ponds. *T.* "Nakaku" and *T.* "Mbita" look very similar in their basic body color. However, T. "Mbita" is distinguishable by a yellow blotch on its flank. *Tropheus* "Mbita" males were crossed with *Tropheus* "Nakaku" females and the offspring was analyzed. Therefore, we used six microsatellites in order to determine reproductive success involved individuals.

6

3. Material and Methods

3.1. Study area and sampling

The wild specimens used in this experiment were sampled from two locations in the southern basin of Lake Tanganyika in Zambia (Fig. 1). The two sampling sites, Nakaku and Mbita, are separated by 50km of shore line. For the breeding experiment ponds with dimensions of 2 x 5 x 0.8m were used and stocked with individuals from two different populations of *Tropheus moorii*: 75 females from Nakaku (08°38´S, 30°52´E) were st ocked and 25 males from Mbita island (08°44´S, 31°06´E). .

The experiment began in March 2005. The pond was equipped with five stone clusters in order to provide hiding and grazing space for fish and territories for males. Before the fish were released into the pond a finclip was taken from each individual and stored in 99% alcohol. In 2006 (after 1,5 years) the parental fish were removed from the pond, transported to Graz and 147 individuals of F1 offspring were counted, to be set back to the pond. In 2007 the F1 fish were sampled for the second time counting 161 individuals which already indicates F2-offspring. Finally, in 2008 in total 133 adult F1 Hybrids were sampled, sexed, scanned for morphology and a fin-clip was taken for genetic analysis. The samples analyzed in this study were taken in 2008.

3.2. Microsatellites

In this study six di-nucleotide microsatellites were used (Table 1). For genotyping five multiplex PCRs were established: multiplex 1 (UNH130 (Lee and Kocher, 1995) and Pzeb3 (Van Oppen *et al.* 1997)), multiplex 2 (Pzeb2 (Van Oppen *et al.* 1997)) and multiplex 3 (UME002 (Parker and Kornfield, 1996), UME003 (Parker and Kornfield, 1996) and TmoM27 (Zardoya *et al.* 1996) with forward primers being labeled with two colors: HEX (UNH130, Pzeb2, UME003 and TmoM27) and 6-FAM (Pzeb3 and UME002). Furthermore, multiplex 4 (UNH130, Pzeb3, TmoM27) and multiplex 5 (UME002, UME003, Pzeb2) were established wherein Pzeb2 forward primer was labeled with NED.

Primer name	Dye	Sequence	References
UME002_F	6-fam	5' tca gag tgc aat gag aca tga at	Parker and Kornfield, 1995
UME002_R	6-fam	5' aat tta gaa gca gaa aat tag acg	Parker and Kornfield, 1995
UME003_F	hex	5' gcc aca tgt aat cat cat act gc	Parker and Kornfield, 1995
UME003_R	hex	5' gag att ttt ttt ggt tcc gtt g	Parker and Kornfield, 1995
Unh130_F	hex	5' agg aag aat agc atg tag caa gta	Lee and Kocher, 1995
Unh130_R	hex	5' gtg tga taa ata aag agg cag aaa	Lee and Kocher, 1995
Pzeb2_F	hex (ned)	5' ttc ggt aga ctg atg ctt tca ta	Van Oppen *et al.*, 1997
Pzeb2_R	hex (ned)	5' aaa gcc aaa ggg tgt gaa ctg a	Van Oppen *et al.*, 1997
Pzeb3_F	6-fam	5' gag cct gca aac ctt act gta aa	Van Oppen *et al.*, 1997
Pzeb3_R	6-fam	5' aag cta cac aaa ttc cac tca ta	Van Oppen *et al.*, 1997
TmoM27_F	hex	5' agg cag gca att acc ttg atg tt	Zardoya *et al.*, 1996
TmoM27_R	hex	5' tac taa ctc tga aag aac ctg tga t	Zardoya et al., 1996

Table 1: List of primers used for genetic analyses

3.3. Molecular genetic methods

3.3.1. DNA extraction

For parental analyses nuclear DNA was extracted from 99% ethanol preserved finclips by ammonium-acetate (AAC) protocol based on Koch (2004) and Chelex extraction protocol based on Walsh PS (1991)(see appendix). In the first case, proteinase K was used for digestion of tissue, followed by ammonium acetate and isopropanol precipitation (Sambrook, *et al.* 1989). Chelex procedure however, is simple, rapid and no digestion enzyme is needed (Walsh, *et al.* 1991).

3.3.2. PCR – conditions

Amplification of microsatellites (multiplex 1 – 3) was carried out in a total volume of 20µl with standard PCR cocktail containing: 13.3 µl aqua dest., 0.5 µl primer (1:10 diluted) (forward and reverse) (10pmol/µl), 0.5 µl dNTP's, 0.5 µl BSA, 2 µl MgCl$_2$ buffer (15 mM), 0.2 µl TAQ polymerase (5 U/µl) and 2.5 µl DNA AAC extraction (1:30 dilution). In case of Chelex extraction 10 µl of DNA extraction (no dilution) were used and PCR cocktail mentioned above was adjusted to 10 µl of rest reaction volume.

Additionally, PCR amplification (microsatellites in multiplex 4 and 5) using a master mix from a commercial source Type-it Microsatellite PCR KIT (Qiagen) (see appendix) was carried out in total volume of 6 µl. This modified PCR cocktail contained: 2.5 µl reaction mix (3mM Mg^{2+}), 0.16 µl primer mix (0.02 µl of each primer concentrations (forward and reversed)), 2.34 µl RNase-free water and 1 µl of DNA extraction.

PCR started with a denaturation phase at 94°C for 3min, followed by 45 cycles of 94°C, 30s; 55°C, 30s; 72°C, 1min and ende d by the final phase of 72°C, 7min.

3.3.3. Capillary electrophoresis

PCR products were prepared for the final analysis in a total volume of 11.5 µl in a cocktail containing: 10 µl Hi – Di (Formamid) and 0,3 µl size standard (Genescan ROX500). 10 µl of this mix and 1.5 of PCR product were denaturated at 95°C for 5min and loaded on an Appli ed Biosystems 3130xl sequencer.

3.3.4. Genotyping with Gene Mapper

Scoring of alleles was performed manually using ABI GeneMapper version 3.7 software (Applied Biosystems, Foster City, CA, USA). An individual strategy was developed to gain equal scoring for all loci. To avoid scoring errors all loci were checked up to ten times. The results of two different PCR methods showed dislocations of the same alleles for 1 – 2 bp in different loci. Therefore, all scoring results were adjusted to scoring results of Type-it Microsatellite PCR KIT.

Allele scoring of the same 6 microsatellites of parental population (Nakaku and Mbita) already existed from previous study. However, the samples were extracted, amplified and anlayzed by sequencer again under current conditions to check potential differences which may have occur using different chemicals and buffer three years ago.

3.3.5. Parental analyses with CERVUS

Parental analyses were carried out using the software program CERVUS 3.0 (Kalinowski *et al.* 2007). This program is normally used to assess pedigree relationships within wild population samples (Marshall *et al.* 1998; Kalinowski *et al.* 2007). However, in this study all parents were known. Therefore, the exclusion method (all parents known) was used which delivers parent - offspring pairs in 100% confidence.

3.3.6. Data analysis

All three populations (Nakaku, $n = 25$; Mbita, $n = 75$ and F1, $n = 133$) were tested for Hardy – Weinberg Equilibrium, observed (Ho) and expected (He) heterozygosity in Arlequin version 3.11 (Excoffier *et al.* 2005) and furthermore, for linkage disequilibrium, genetic diversity and number of sampled alleles in Fstat 2.9.3.2. (Goudet, 2002).

4. Results

4.1. Allele scoring

Unfortunately, microsatellites of almost half of the samples could not be successfully amplified by our standard PCR microsatellite protocol (data not shown). An alternative amplification protocol using the Type-it Microsatellite PCR KIT successfully amplified microsatellites in most of the problematic samples.

After comparison of old and new parental population scorings (data not shown), three microsatellites (Pzeb3, UME003 and Unh130) showed identical scorings and the other three (UME002, Pzeb2, TmoM27) loci showed 1 - 2 bp displacement. In these cases, parental scoring data was adjusted according to the new results based on the Type-it Microsatellite PCR KIT. In this way, a homogeneous data set was achieved for the parental and F1 - offspring population.

Unfortunately, peaks of the Pzeb2 locus in the F1 offspring generation (data not shown) were somewhat irregularly shifting, therefore, it was sometimes impossible to distinguish between stutter bands and alleles. Consequently, this locus was assessed manually, and only included when five loci were not sufficient for parent – pair identification.

4.2. Molecular analysis

Gene diversity (Table 2) decreased in the F1 population, as expected. Interestingly, TmoM27 showed slightly higher gene diversity than the parental population. On the other hand, Pzeb3 and Ume002 showed lower heterozygosity in Nakaku females and in Ume002 in F1 population.

To my expectations, F1 hybrid population showed a significant deviation from Hardy – Weinberg equilibrium, while both parental populations were in Hardy Weinberg Equilibrium (data not shown). Our test for linkage disequilibrium showed no significant results (data not shown).

Gene diversity per locus per population	F1 - Offsprings	Mbita males	Nakaku females
Pzeb2	0.913	0.938	0.933
Pzeb3	0.717	0.768	0.649
UNH130	0.899	0.855	0.93
UME002	0.647	0.745	0.688
UME003	0.908	0.943	0.923
TmoM27	0.859	0.84	0.714

Table 2: Gene diversity in parental and F1 population

4.3. Parental analysis with CERVUS

The results of the parental analysis in the parent and offspring generation are based on complete concordance of all scores of alleles in five to six loci. Therefore, the confidence of identified parent – offspring pairs is 100%. In this study 82% (109 offspring) were successfully assigned to their parents. 18% (24 offspring individuals) showed mismatching genotypes. Furthermore, within the F1 generation 4.5% (6 fish) were identified as F2 generation individuals, however, only four could be successfully assigned to their parents from F1 population. Moreover, the analysis found two fish (1.5%) where the mother could be identified but the father was not found.

A total of 137 alleles were found for the six microsatellites (Fig. 2). Fstat showed similar numbers of alleles in all loci for the F1 generation as well as for both parental populations. However, the 25 Mbita males had generaly less alleles than the 75 Nakaku females. Interestingly, the F1 generation had the same number of alleles or even more than the Nakaku females, except for Ume003 which showed lower number of alleles in the F1 generation.

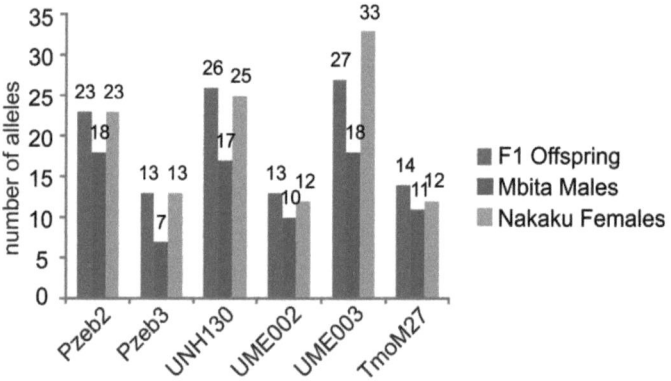

Fig 2. Number of alleles per locus

14

The results based on 82% of the individuals of the F1 hybrid population indicated reproductive success for 18 Mbita males (72%) and 43 Nakaku females (57%), as shown in Figure 3. Two males were particularly successful with 30 (M11) and 22 (M2) surviving offspring, respectively. Two most successful females (F2 and F35), however, succeeded with 8 surviving offspring. Furthermore, three females (F4, F54, F59) were also moderately successful with 5 surviving offspring.

Surprisingly, our results showed no correlation between body size and number of offspring for neither male nor female parents, with $R^2 = 0.0275$ for Mbita males and $R^2 = 0.0553$ for Nakaku females (Fig. 4).

Clearly, the number of matings differed dramatically in both parental males and females (Fig. 5). The two most successful Mbita males mated with 10 and 18 females, respectively, while the others managed to mate with one to five females only. The majority of the Nakaku females mated with only one partner, while 15 females chose two mates, 2 females chose three and four partners, respectively, and surprisingly one female produced offspring with five males.

Furthermore, results showed a strong positive correlation between the number of mates and the number of offspring for both parents (Fig. 6).

Parental analysis identified 20 full sibling – groups (see Appendix, Table 3): one with 7, one with 4, five with 3 and thirteen with 2 siblings. There were also several half siblings, as evident from Table 3: For example, M2 male and F2 female have 7 full sibling offspring and these 7 offspring are half siblings with all the other individuals on the vertical and horizontal axis of M2 and F2 in Table 3. Values marked with pink in M11 and grey in M13 vertical axis represent one individual each with two possible mothers (marked with the same color as potential offspring). The analysis based on five loci could in both cases identify the father, however, for both individuals (pink and grey) two possible mothers emerged. Unfortunately, in this case the sixth locus, Pzeb2, had missing data and therefore, the real mother could not be determined.

The initial analysis revealed 103 (77%) true parent – offspring pairs of F1 generation. In the second round of analysis, 30 fish (23%), which could not be assigned to their parents, were tested against rest of 103 fish (77%) as potential parents. Interestingly, six of these fish were found as potential F2 generation offspring. With 100% confidence four (3%) parent – offspring pairs of F2 were identified and for the other two fish (2%) either the father or the mother could not be assigned. One fish remained with two possible mothers and another one with two possible fathers. Additionally, Fig. 7 represents body size distribution of parents from F1 population of only four F2 individuals which could be assigned to their parents.

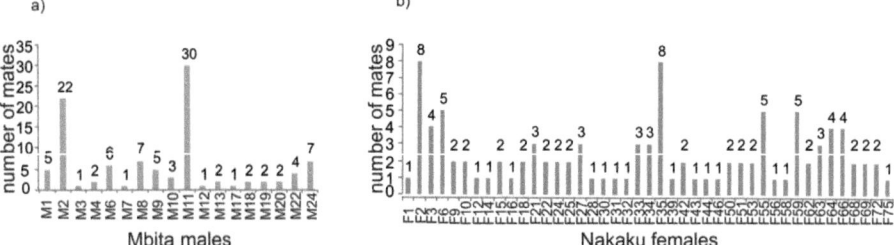

Fig 3. Reproductive success for Mbita males (a) and Nakaku females (b)

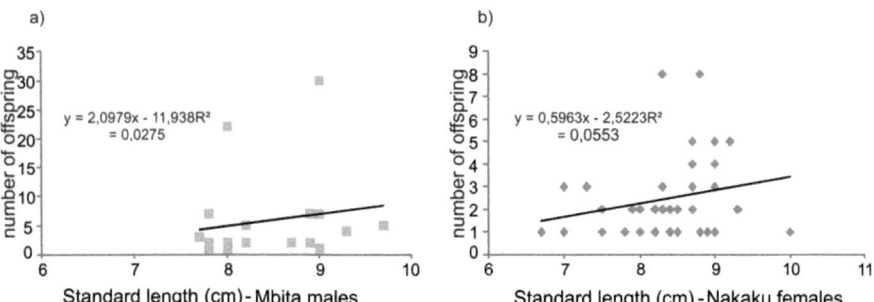

Fig 4. Reproductive success of Mbita males (a) and Nakaku females (b) in relation to standard length

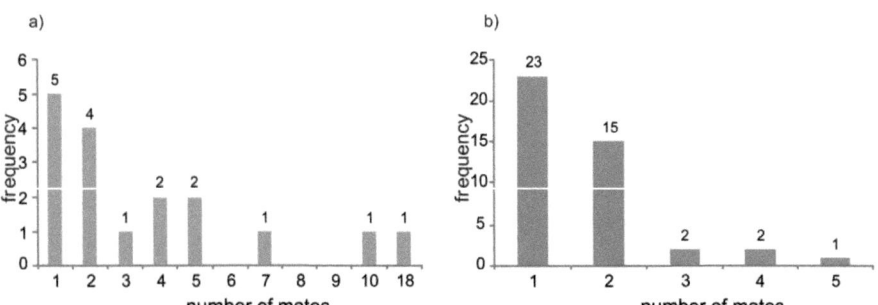

Fig 5. Number of females for Mbita males (a) and Nakaku females (b).

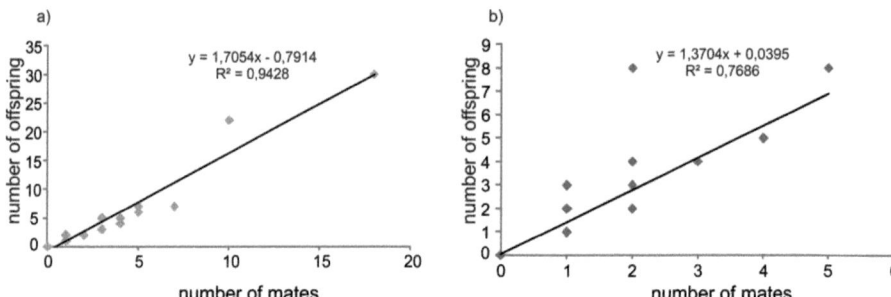

Fig. 6. Correlation between number of offspring and number of mates for Mbita males (a) and Nakaku females (b)

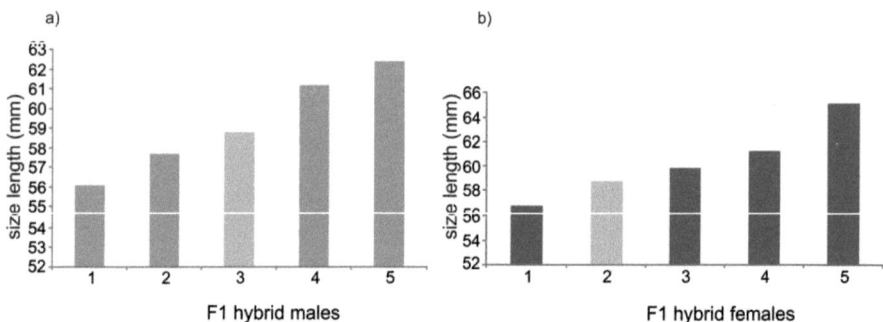

Fig. 7. Size of F1 hybrid parents for males (a) and females (b) with median for 4 true parent - offspring pairs.

5. Discussion

5.1 Methodological aspects

Allele scoring

The success in the amplification of microsatellites can be influenced by many factors. Clearly, good DNA quality is a prerequisite. Although samples used in this study were stored for three years in 99% analysis garde ethanol, frequent amplification problems arose. The DNA - extraction according to the AAC - protocol (Koch, 2004) suddenly resulted in low DNA yields (1.5 - 4 ng), which was suspected to be the reason for unsuccessful amplification of almost half of the samples using our standard PCR - protocol. An alternative DNA - extractions based upon Chelex 100 (Walsh PS, 1991) and the modification of our amplification reagents including Type-it microsatellite PCR KIT yielded better results. We suspect that the problems are associated with the use of new vials with a screw - cup and black sealings which turned out not to be ethanol - resistant. Older samples took on dark color by solving a substance from the sealing ring. This substance might have acted as inhibitor and caused problems for amplification. By including the Type-it microsatellite PCR KIT we managed to amplify most of the samples. We also had to overcome scoring shifts, but eventually, all allele scorings of parental and F1 hybrid population could be adjusted to the scoring results with the Type-it microsatellite PCR KIT into a homogeneous data set.

Molecular analysis

As expected, the test for Hardy – Weinberg equilibrium resulted, apart from both parental populations, in a deviation in the F1 hybrid generation. This was expected, because the F1 hybrids consist of several full and half siblings, and therefore, cannot be in equilibrium.

5.2. Biological aspects

Male - male competition turned out to be rather low among the Mbita males, given that 72% (18) managed to produce surviving offspring. Fourteen out of the 18 successfully reproducing males had more than one surviving offspring. Supposedly, unassigned 7 males might have been the parents of the other 24 unresolved offspring individuals. However, they might as well have failed to reproduce, because they could not gain a sizable territory which is most likely a crucial mating advantage.

On the other hand, competition among females seems to be rather high, given that only 57% (43) were successfully producing surviving offspring. Within 32 unassigned females, at least 18 remained without offspring. Presumably, their entire offspring died in some of these females, and a few already sampled females may have died before they managed to reproduce. Other female individuals may not have been able to reproduce due to stress factors resulting from the high population density when compared to natural circumstances.

It seems plausible that males try to mate with as much females as possible (Bateman, 1948). A much stronger positive correlation between the number of mates and number of offspring was observed for Mbita males, implying that males are exposed to a higher selective pressure. Such behavior can be explained by intrasexual selection (male-male competition) or intersexual selection by female choice (Egger et al. 2006; Egger et al. 2008; Sefc, 2008).

Female preference for large males has been observed in many fish species (Ptacek & Travis, 1997; Rosenthal & Evans, 1998; Basolo, 2004; Hudman & Gotelli, 2007). However, the present study shows no such preference like already observed in T.moorii (Egger et al. 2010). The two males with the most offspring differed dramatically in body size. With a standard length of 8 cm,

M2 male belonged to the smaller males and M11, with its 9 cm standard length, to the largest males. Indeed, it has been observed that females preferred more active males rather than large ones (Steinwender et al. 2012). Presumably, the most successful male M2 was dominant and highly active or aggressive, as opposed to other large males with distinctly less offspring, and even to the largest male which had only five surviving offspring individuals. Both most successful males could have as well occupied a corner territory in the pond which is easier to defend as opposed to a territory in the middle.

Many studies in other fish species documented female preference for more active males (Bischoff et al. 1985; Knapp & Kovach, 1991; Carvalho et al. 2003; Maan et al. 2004; Reichard et al. 2005), the mallard (Bossema & Kruijt, 1982; Kruijt et al. 1982), and a wolf spider (Shamble et al. 2009). It has been suggested that male courtship activity could be a consequence of female attention (Takahashi et al. 2008). Other types of female communication such as olfactory signals seem also possible, through which males become motivated to display a higher courtship activity (Steinwender et al. 2012). In many fish species particular movements in courtship were observed to produce other signals. Several cichlid species including T. moorii, exhibit a part of courtship movement being associated with the production of sound (Nelissen, 1978; Amorim et al. 2008; Smith & van Staaden, 2009), and this acoustic signal was suggested to influence the female preference in Lake Victoria cichlids (Verzijden et al. 2010). In addition, through courtship movements odors may be dispersed which could as well support olfactory signaling (Blais et al. 2007; Agbail et al. 2010).

Females also showed no correlation between body size and fitness. One of both most successful females was medium size and the second successful female was larger. However, all three females with 5 offspring belonged to larger individuals. In our data set, no trend can be seen and such reproductive success can result as a coincidence.

Parental analysis revealed 20 full - sibling groups and 43 families with one offspring. Within these families 15 females had only one offspring which indicates that many individuals bred out but most of their offspring did not survive. Presumably, several of the more successful females had originally more offspring as well. One might speculate that the survival in the pond environment is influenced by genetic factors, on which natural selection acts. It is thus possible that the most successful parents have genes that are better adapted to the pond environment and pond circumstances than other alleles.

Several previous studies showed that normally, *T. moorii* broods are sired by a single male (Egger *et al.* 2006). Male and female form a temporary pair in the territory of the male. Surprisingly, one female mated with 5 different males which implies to either unexpected five generations of this female in 1.5 years or a possible sneaking in a pond being more likely due to high fish density. In other related Tropheini sneaking males have been observed mixing among spawning pairs such as in *Pseudosimochromis curvifrons* and *Simochromis diagramma* (Kuwamura 1987), *Ctenochromis horei* (Ochi 1993*a*) and *Gnathochromis pfefferi* (Ochi 1993*b*). Unfortunately, the number of broods in this study could not be determined. All F1 offspring were full grown between 4.8 and 6.6 cm, therefore it was impossible to distinguish between broods. Therefore, it is as well possible that a female spawned voluntarily with more than one male under such artificial high-density conditions.

Within the F1 hybrid population, six fish were discovered to represent the F2 generation, however only four of these could actually be assigned to their true parents, while for the two other F2-fish either mother or father could not be assigned. Presumably, some F1 or F2 fish were genetically so similar that the assigned parents could actually be their full siblings. Surely, we can speculate

23

that some of the unresolved 24 fish also consist of F2 generation. Presumably, this data set still consists of some unavoidable scoring errors, or particular parent – offspring pairs could not be found for other reasons. Possibly, some of the female fish were brought into the pond while they already mouthbrooded and consequently, data for the potential fathers would be missing. Theoretically, backcrossing between F1 (first generation being sexually mature after one year) and Back-up parents might occur as well, however this test was not carried out.

Steinwender *et al.* (2012) mentions a simulation on *Tropheus* reproduction suggesting that pair bonding and monogamous spawning must not curb the potential for sexual selection, if female mate choice determines the variance in male reproductive success (Sefc, 2008). This study supports this prediction, as two males clearly had significantly more matings than the others. On the other hand, monogamous spawning cannot be confirmed in our hybrid crossing experiment. However, as argued above, it seems that male sneaking or multiple male visits of females during spawning are more likely in a pond due to artificially high fish density.

6. References

Amorim, M. C. P., Simoes, J. M. , Fonseca, P. J. & Turner, G. F. 2008. Species differences in courtship acoustic signals among five Lake Malawi cichlid species (*Pseudotropheus spp.*). Journal of Fish Biology **72:** 1355–1368.

Arnegard, M. E. & Kondrashov, A. S. 2004. Sympatric speciation by sexual selection alone is unlikely. Evolution 58: 222–237.

Basolo, A. L. 2004. Variation between and within the sexes in body size preferences. Animal Behaviour 68: 75–82.

Bateman, A.J. 1948. Intrasexual selection in *Drosophila*. Heredity **2:** 349-368.

Bischoff, R. J., Gould, J. L. & Rubenstein, D. I. 1985. Tail size and female choice in the guppy (Poecilia reticulata). Behavioral Ecology and Sociobiology 17: 253–255.

Bossema, I. & Kruijt, J. P. 1982. Male Activity and female mate acceptance in the mallard (Anas platyrhynchos). Behaviour 79: 313–324.

Carvalho, N., Afonso, P. & Serrano Santos, R. 2003. The haremic mating system and mate choice in the wide - eyed flounder, *Bothus podas*. Environmental Biology of Fishes 66: 249–258.

Coulter, G.W. (Ed.) 1991. Lake Tanganyika and its Life. Oxford University Press, New York.

Coyne, J. A. & Orr, H. A. 2004. Speciation. Sunderland, MA: Sinauer Associates.

Danley, P. & Kocher, T. 2001. Speciation in rapidly diverging systems: lessons from Lake Malawi. Mol. Ecol. 10: 1075–1086.

Egger, B., Obermüller, B., Phiri, H., Sturmbauer, C., Sefc, K.M. 2006. Monogamy in the maternally mouthbrooding Lake Tanganyika cichlid fish *Tropheus moorii*. Proceedings of the Royal Society B: Biological Sciences 273: 1797–1802.

Egger, B., Koblmueller, S., Sturmbauer, C., Sefc, K.M. 2007. Nuclear and mitochondrial data reveal different evolutionary processes in the Lake Tanganyika cichlid genus Tropheus. Bmc Evolutionary Biology 7: 137.

Egger, B., Obermüller, B., Eigner, E., Sturmbauer, C., Sefc, K.M. 2008. Assortative mating preferences between color morphs of the endemic Lake Tanganyika cichlid genus *Tropheus moorii*. Hydrobiologia 615: 37-48.

Egger, B., Mattersdorfer, K., Sefc, M. K. 2010. Variable discrimination and asymmetric preferences in laboratory tests of reproductive isolation between cichlid color morphs. J. Evol. Biol. 23: 433 – 439.

Excoffier, L. Laval, G., Schneider, S. 2006. Arlequin version 2.11: An integrated Software Package for population genetics data analysis, Computational and molecular population genetics Lab (CMPG). Institute of Zoology University of Berne, Switzerland. Available from: http://cmpg.unibe.ch/software/arlequin3

Fryer, G., Iles, Oliver, T.D. & Boyd 1972. The cichlid fishes of the Great Lakes of Africa: their biology and evolution, Edinburgh.

Goudet, J. 2002. Fstat 2.9.3.2., a program to estimate and test gene diversity and fixation indices. Available from http://www2.unil.ch/popgen/softwares/fstat.htm

Greenwood, P. H. 1984. African cichlids and evolutionary theories. A. A. Echelle and I. Kornfield, eds. Evolution of fish species flocks. University of Maine at Orono Press, Orono, Maine; 141–154.

Hudman, S. P. & Gotelli, N. J. 2007. Intra- and intersexual selection on male body size are complimentary in the fathead minnow (*Pimephales promelas*). Behaviour 144: 1065–1086.

Kalinowski, S.T., Taper, M.L., Marshall, T.C. 2007. Revising how the computer program Cervus accommodates genotyping error increases success in paternity assignment. Molecular Ecology, 16: 1099–1106.

Knapp, R. A. & Kovach, J. T. 1991. Courtship as an honest indicator of male parental quality in the bicolor damselfish, *Stegastes partitus*. Behavioral Ecology 2: 295–300.

Knight, M. E. & Turner, G. F. 2004. Laboratory mating trials indicate incipient speciation by sexual selection among populations of the cichlid fish *Pseudotropheus zebra* from Lake Malawi. Proc. R. Soc. B 271: 675–680.

Koblmüller, S., Sefc, K.M., Sturmbauer, C. 2008a. The Lake Tanganyika cichlid species assemblage: recent advances in molecular phylogenetics. Hydrobiologia 615: 5–20.

Koblmüller, S., Salzburger, W., Obermüller, B., Eigner, E., Sturmbauer, C., Sefc, M. K. 2011. Seperated by sand, fused by dropping water: habitat barriers and fluctuating water levels steer the evolution of rock dwelling cichlid populations in Lake Tanganyika. Mol. Ecol. 20: 2272-2290.

Koch, M. 2004. Population structure and intraspecific genetic variation of *Tylochromis polylepis*, a cichlid species of Lake Tanganyika. Diplomarbeit der Naturwissenschaften, Karl – Franzens Universität Graz, A.

Koch, M., Hadfield, D. Jarod, Sefc, M. K., Sturmbauer, C. 2008. Pedigree reconstruction in wild cichlid fish populations. Molecular Ecology 17: 4500 – 4511.

Kocher, T.D. 2004. Adaptive evolution and explosive speciation: the cichlid fish model. Nat. Rev. Genet. 5: 288–298.

Kornfield, I., Smith, P.F. 2000. African cichlid fishes: model systems for evolutionary biology. Annu. Rev. Ecol. Syst. 31: 163–196.

Kuwamura, T. 1987. Male mating territory and sneaking in a maternal mouthbrooder, *Pseudosimochromis curvifrons* (Pisces; Cichlidae). J. Ethol. 5: 203–206.

Lee, W.J., Kocher, T.D. 1995. Microsatellite DNA markers for genetic mapping in *Oreochromis niloticus*, J Fish Biol. 49: 169–171.

Lipinski, M. J. ,Amigues, Y., Blasi, M., Broad, T. E., Cherbonnel, C., Cho, G. J., Corley, S., Daftari, P., Delattre, D. R., Dileanis, S., Flynn, J. M., Grattapaglia, D., Guthrie, A., Harper, C., Karttunen, P. L., Kimura, H., Lewis, G. M., Longeri, M., Meriaux, J.-C., Morita, M., Morrin-O'Donnell, R. C., Niini, T., Pedersen, N. C., Perrotta, G., Polli, M., Rittler, S., Schubbert, R., Strillacci, M. G., Van Haeringen, H., Van Haeringen W., and Lyons I. A. 2007. An international parentage and identification panel for the domestic cat (*Felis catus*), International Society for Animal Genetics, Animal Genetics, 38: 371–377.

Maan, M. M., Seehausen, O., Söderberg, L., Johnson, L., Ripmeester, E. A. P., Mrosso, H. D. J., Taylor, M. I., van Dooren, T. J. M. & van Alphen, J. J. M. 2004. Intraspecific sexual selection on a speciation trait, male coloration, in the Lake Victoria cichlid *Pundamilia nyererei*. Proc. R. Soc. B 271: 2445–2452.

Marshall, T.C., Slate, J., Kruuk, L.E.B., Pemberton, J.M. 1998. Statistical confidence for likelihood-based paternity inference in natural populations. Molecular Ecology, 7: 639–655.

Meyer A, Kocher T.D., Basasibwaki P, Wilson A.C. 1990. Monophyletic origin of Lake Victoria cichlid fishes suggested by mitochondrial DNA sequences. Nature 347: 550–553.

Nelissen, M. 1978. Sound production by some Tanganyikan cichlid fishes and a hypothesis for the evolution of their communication mechanisms. Behaviour 64: 137–147.

Ochi, H. 1993a. Mate monopolization by a dominant male in a multi-male social group of a mouthbrooding cichlid, *Ctenochromis horei*. Jpn. J. Ichthyol. 40: 209–218.

Ochi, H. 1993b. Maintenance of separate territories for mating and feeding by males of a maternal mouthbrooding cichlid, *Gnathochromis pfefferi*, in Lake Tanganyika. Jpn. J. Ichthyol. 40: 173–182.

Parker A., Kornfield I. 1996. Polygynandry in *Pseudotropheus zebra*, a cichlid fish from Lake Malawi. Env. Biol. Fish. 47: 345–352.

Pauers, M. J., McKinnon, J. S. & Ehlinger, T. J. 2004. Directional sexual selection on chroma and within - pattern color contrast in *Labeotropheus fuelleborni*. Proc. R. Soc. B 271: (Suppl. 6), 444–447.

Poll, M. 1986. Classification des Cichlidae du lac Tanganika: tribus, genres et espèces. Acad. R. Belg. Mem. Cl. Sci. 45: 1–163.

Ptacek, M. B. & Travis J. 1997. Mate choice in the sailfin molly, *Poecilia latipinna*. Evolution 51: 1217–1231.

Reichard, M., Bryja, J., Ondrackova, M., Kaniewska, P. & Smith, C. 2005. Sexual selection for male dominance reduces opportunities for female mate choice in the European bitterling (*Rhodeus sericeus*). Mol. Ecol. 14: 1533–1542.

Rosenthal, G. G. & Evans, C. S. 1998. Female preference for swords in *Xiphophorus helleri* reflects a bias for large apparent size. Proceedings of the National Academy of Sciences United States of America 95: 4431–4436.

Rossiter, A., 1995. The cichlid fish assemblages of Lake Tanganyika: ecology, behavior and evolution of its species flock. Advances in ecological research 26: 187-252

Salzburger, W., Meyer, A. 2004. The species flocks of East African cichlid fishes: recent advances in molecular phylogenetics and population genetics. Naturwissenschaften 91: 277–290.

Salzburger, W., Niederstätter, H., Branfstätter, A., Berger, B., Parson, W., Snoeks, J., Sturmbauer, C. 2006. Color – assortative mating among populations of *Tropheus moorii*, a cichlid fish from Lake Tanganyika, East Afrika. Proc Biol Sci. 273(1584): 257–266.

Sambrook, J., Fritsch E.F., Maniatis T. 1989. Molecular Cloning: A Laboratory Manual, 2nd edn. Cold Spring Harbor Laboratory Press, New York.

Schupke, P. 2003. African cichlids II: Tanganyika I: Tropheus. Germany: Aqualog, A.C. S Gmbh.

Seehausen, O. & van Alphen, J. J. M. 1999. Can sympatric speciation by disruptive sexual selection explain rapid evolution of cichlid diversity in Lake Victoria? Ecol. Lett. 2: 262–271.

Seehausen, O., Witte, F., Van Alphen, J. J. M. & Bouton, N. 1998. Direct mate choice maintains diversity among sympatric cichlids in Lake Victoria. J. Fish Biol. 53: 37–55.

Sefc, K.M. 2008. Variance in reproductive success and the opportunity for selection in a serially monogamous species: simulations of the mating system of *Tropheus* (Teleostei: Cichlidae). Hydrobiologia 615: 21–35

Sefc, K. M. 2011. Mating and parental care in Lake Tanganyika's cihlids. International journal of evolutionary biology 2011: 1-20

Shamble, P. S., D. J. Wilger, K. A. Swoboda & Hebets, E. A. 2009. Courtship effort is a better predictor of mating success than ornamentation for male wolf spiders. Behavioral Ecology 20: 1242–1251.

Smith, A. R. & van Staaden, M. J. 2009. The association of visual and acoustic courtship behaviors in African cichlid fishes. Marine and Freshwater Behaviour and Physiology 42: 211–216.

Steinwender, B., Koblmüller, S., Sefc, K.M. 2012. Concordant female mate preferences in the cichlid fish *Tropheus moorii*. Hydrobiologia 682: 121-130

Stiassny, M.L.J, Meyer A. 1999. Cichlids of the Rift Lakes. Sci. Am. 280: 64–69.

Sturmbauer, C. & Dallinger, R. 1995. Diurnal variation of spacing and foraging behaviour in *Tropheus moorii* (Cichlidae) in Lake Tanganyika, Eastern Africa. Netherlands Journal of Zoology no 3-4, 45: 386–401.

Sturmbauer, C. 1998. Explosive speciation in cichlid fishes of the African Great Lakes: a dynamic model of adaptive radiation. J. Fish Biol. 53: 18–36.

Sturmbauer, C. 2000. Die Seen Ostafrikas und ihre Buntbarsche. Biologie in unserer Zeit 6: 354-366

Sturmbauer, C.; Husemann, M., and Danley, P. D. 2011. Explosive speciation and adaptive radiation of East African Cichlid Fishes. Biodiversity Hotspots: Distribution and Protection of Conservation Priority Areas, F. E. Zachos and J. C. Habel, Springer, Eds., pp. 333–362,

Takahashi, T. 2003. Systematics of Tanganyikan cichlid fishes (Teleostei: Perciformes). Ichthyol. Res. 50: 367–382.

Van Oppen, M.J.H., Rico, C., Deutsch, T.C., Turner, G.F., Hewitt, G.M. 1997. Isolation and characterization of microsatellite loci in the cichlid fish *Pseudotropheus zebra*. Mol. Ecol. 6: 387–388

Verheyen, E., Salzburger W., Snoeks J., Meyer A. 2003. Origin of the superflock of cichlid fishes from Lake Victoria, East Africa. Science 300: 325–329.

Verzijden, M. N., J. van Heusden, N., Bouton, F., Witte, C., Slabbekoorn Cate & H. 2010. Sounds of male Lake Victoria cichlids vary within and between species and affect female mate preferences. Behavioral Ecology 21: 548–555.

Walsh, P.S., Metzger, D.A., Higuchi, R. 1991. Chelex 100 as a medium for simple extraction of DNA and PCR – based typing from forensic material, Biotechniques, 10: 506-513

Wiedl, T. 2008. Reproductive success and sib-hip reconstruction in two pond-bred populations of the Lake Tanganyika cichlid fish species *Tropheus moorii*. Magisterarbeit, Karl-Franzens Universität Graz, A.

Yanagisawa, Y. & Nishida, M. 1991. The social and mating system of the maternal mouthbrooder *Tropheus moorii* (Cichlidae) in Lake Tanganyika. Japanese Journal of Ichthyology 38: 3, 271–282.

Yanagisawa, Y. and Sato, T. 1990. Active browsing by mouthbrooding females of *Tropheus duboisi* and *Tropheus moorii* (Cichlidae) to feed the young and/or themselves. Environmental Biology of Fishes 27: 1, 43–50.

Zardoya, R.; Vollmer, D.M.; Craddock, C.; Streelman, J.T.; Karl, S.A.; Meyer, A. 1996. Evolutionary conservation of microsatellite flanking regions and their use in resolving the phylogeny of cichlid fishes (Pisces: Perciformes). Proc R Soc 263: 1589–1598

A. Appendix

A.1. AAC extraction protocol

<u>Ammonium Acetat Exktraktion (AAC) for multichannel purpose (48 or 96 samples)</u>

1) Put samples in 400 µL extraction buffer (320 µl buffer + 80 µl 10% SDS) and add 5 – 5,5 µl proteinase K. Add tissue and mix it well.

2) Incubate samples at 45°C and 450 rpm in thermocy cler. After one hour mix strips until the whole tissue dissolves. Keep strips in thermocycler over night.

3) Next day put strips after short centrifugation into refrigerator (-20°C) until they are completely frozen.

4) After defrosting add 230 µl cold AAC (eliminates protein) and mix it well.

5) Centrifuge for 40 – 45 minutes.

6) Pipette supernatant into a new tube carefully, not to lose the pellet.

7) Add 400 µl cold isopropanol and mix it well.

8) Put it on ice over night at -20°C.

9) Centrifuge for 40 – 45 minutes.

10) Remove the supernatant (isopropanol) without losing the pellet.

11) Washing-step 1: Add 400µl 70% ethanol and mix gently.

12) Centrifuge for 20 minutes.

13) Remove the supernatant (ethanol).

14) Washing-step 2: Add 400µl 100% ethanol.

15) Centrifuge for 20 minutes.

16) Remove the supernatant (ethanol).

17) Dry pellet at 60°C for 30 minutes.

18) Add 60µl TE – buffer.

19) Place samples into thermocycler at 37°C over ni ght to bring DNA into solution.

20) Put samples into refrigerator.

A.2. Chelex extraction protocol

Chelex Extraktion for 10% solution Walsh PS (1991)

1) Weight 50mg Chelex 100 for each sample.

2) Add 500 µl ddH2O.

3) Add very small piece of tissue.

4) Vortex two times for 20s.

5) Centrifuge shortly by high speed.

6) Incubate on 450rpm for 20min at 95°.

7) Vortex for 10 – 15s.

8) Spin down at high speed - prior to use

A.3. Type-it Microsatellite PCR KIT modified protocol

The KIT contained: Type-it Multiplex PCR Master Mix (with 3 mM Mg^{2+}), RNase-free H$_2$O and Q-solution (for templates with GC-rich regions or complex secondary structure - was not used here). For Primer Mix for 96 samples 1.92 µl of each primer concentration (forward and reversed) were used (0.02 µl per sample).

Type-it Microsatellite PCR KIT for 96 Samples

PCR: 5µl MM (Type-it Multiplex PCR Master Mix+ RNase-free H$_2$O+Primer Mix) + 1 µl DNA extract

Type-it Multiplex PCR Master Mix	240.00 µl
RNase-free H$_2$O	224.64 µl
Primer Mix	15.36 µl

A.4. Microsatellite PCR standard protocol

Microsatellite PCR cocktails for 96 Samples

PCR: 17.5 µl MM + 2.5 µl DNA extract

Multiplex 1 – Plate (100)

UNH130 forward	50 µl
UNH130 reverse	50 µl
Pzeb3 forward	50 µl
Pzeb3 reverse	50 µl
dNTP's	50 µl
BSA	50 µl
TAQ Puffer	200 µl
TAQ Polymerase	20 µl
H2O	1230 µl

Multiplex 2 – Plate (100)

Pzeb2 forward	50 µl
Pzeb2 reverse	50 µl
dNTP's	50 µl
BSA	50 µl
TAQ Puffer	200 µl
TAQ Polymerase	20 µl
H2O	1330 µl

Multiplex 3 – Plate (100)

TmoM27 forward	50 µl
TmoM27 reverse	50 µl
UME003 forward	50 µl
UME003 reverse	50 µl
UME002 forward	50 µl
UME002 reverse	50 µl
dNTP's	50 µl
BSA	50 µl
TAQ Puffer	200 µl
TAQ Polymerase	20 µl
H2O	1130 µl

	M1	M2	M3	M4	M6	M7	M8	M9	M10	M11	M12	M13	M17	M18	M19	M20	M22	M24	Total of F	SL (cm)
F1										1									1	9
F2		7								1									8	8,3
F3								2		2									4	8,7
F6		1								2		1		1					5	8,7
F9					1							1							2	8,5
F10							1											1	2	8,2
F12													1						1	8,4
F14												1							1	8,5
F15										1								1	2	8,3
F16										1									1	8,9
F18		2																	2	8,2
F21	3																		3	7,3
F22								1		1									2	8,7
F24		1								1		1							2	7,5
F25				2															2	7
F27		3																	3	8,7
F28	1																		1	8
F30		1																	1	10
F31																	1		1	7,8
F32		1																	1	8,2
F33							2								1				3	7,3
F34		3																	3	9
F35						1	1			4					1			1	8	8,8
F39							1												1	8,8
F42				2															2	7,5
F43										1+1									1	7,8
F44					1														1	7
F46																	1		1	6,7
F50										2									2	9,3
F51										2									2	8,3
F53											1						1		2	8
F55					1		2	1										1	5	9
F56		1																	1	8
F58																1			1	8,4
F59		2						1		1				1					5	9,2
F62									1								1		2	8,4
F63										3									3	8,3
F64			1							3									4	8,7
F66	1															2		1	4	9
F68										1								1	2	7,9
F69										2									2	8
F72					1				1										2	8,5
F75									1										1	7,5
Total of M	5	22	1	2	6	1	7	5	3	30	1	2	1	2	2	2	4	7		
SL (cm)	8,2	8	8	8,9	9	9	8,9	9,7	7,7	9	7,8	8,7	8	8,2	8	7,8	9,3	7,8		

A. Table 3. Family table with 43 Nakaku females (F) and 18 Mbita males (M). Full siblings and individuals with most offspring and their standard length are marked with thick letters. Values marked with pink in M11 and grey in M13 vertical axis represent one individual each with two possible mothers (marked with the same colour as potential offspring).

Mbita males	Number	Nakaku Female	Number	Nakaku Female	Number
9381	M1	9406	F1	9449	F44
9382	M2	9407	F2	9450	F45
9383	M3	9408	F3	9451	F46
9384	M4	9409	F4	9452	F47
9385	M5	9410	F5	9453	F48
9386	M6	9411	F6	9454	F49
9387	M7	9412	F7	9455	F50
9388	M8	9413	F8	9456	F51
9389	M9	9414	F9	9457	F52
9390	M10	9415	F10	9458	F53
9391	M11	9416	F11	9459	F54
9392	M12	9417	F12	9460	F55
9393	M13	9418	F13	9461	F56
9394	M14	9419	F14	9462	F57
9395	M15	9420	F15	9463	F58
9396	M16	9421	F16	9464	F59
9397	M17	9422	F17	9465	F60
9398	M18	9423	F18	9466	F61
9399	M19	9424	F19	9467	F62
9400	M20	9425	F20	9468	F63
9401	M21	9426	F21	9469	F64
9402	M22	9427	F22	9470	F65
9403	M23	9428	F23	9471	F66
9404	M24	9429	F24	9472	F67
9405	M25	9430	F25	9473	F68
		9431	F26	9474	F69
		9432	F27	9475	F70
		9433	F28	9476	F71
		9434	F29	9477	F72
		9435	F30	9478	F73
		9436	F31	9479	F74
		9437	F32	9480	F75
		9438	F33		
		9439	F34		
		9440	F35		
		9441	F36		
		9442	F37		
		9443	F38		
		9444	F39		
		9445	F40		
		9446	F41		
		9447	F42		
		9448	F43		

A. Table 4: List of ID numbers of all Mbita males (M1 – M25) and Nakaku females (F1 – F75)

Mbita males	Number	Nakaku Female	Number
9381	M1	9406	F1
9382	M2	9407	F2
9383	M3	9408	F3
9384	M4	9411	F6
9386	M6	9414	F9
9387	M7	9415	F10
9388	M8	9417	F12
9389	M9	9419	F14
9390	M10	9420	F15
9391	M11	9421	F16
9392	M12	9423	F18
9393	M13	9426	F21
9397	M17	9427	F22
9398	M18	9429	F24
9399	M19	9430	F25
9400	M20	9432	F27
9402	M22	9433	F28
9404	M24	9435	F30
		9436	F31
		9437	F32
		9438	F33
		9439	F34
		9440	F35
		9444	F39
		9447	F42
		9448	F43
		9449	F44
		9451	F46
		9455	F50
		9456	F51
		9458	F53
		9460	F55
		9461	F56
		9463	F58
		9464	F59
		9467	F62
		9468	F63
		9469	F64
		9471	F66
		9473	F68
		9474	F69
		9477	F72
		9480	F75

A. Table 5: List of ID numbers of Mbita males (M1 - M24) and Nakaku females (F1 - F75) with reproductive success

A. Table 6: Genotype table of parental populations Mbita males (9381-9405) and Nakaku females

Individuum	Morph	Sex	Pzeb 2	Pzeb 2	Pzeb3	Pzeb3	UNH130	UNH130	UME002	UME002	UME003	UME003	TmoM27	TmoM27
9381	MbMaleXNakakuF	m	265	267	322	330	175	175	235	239	216	222	396	410
9382	MbMaleXNakakuF	m	229	267	330	332	175	177	235	235	218	236	394	398
9383	MbMaleXNakakuF	m	229	251	330	330	175	175	239	235	214	226	390	396
9384	MbMaleXNakakuF	m	241	249	326	330	171	187	233	237	212	218	398	406
9385	MbMaleXNakakuF	m	241	267	322	322	199	227	235	235	230	230	406	410
9386	MbMaleXNakakuF	m	251	275	322	330	175	175	235	239	174	250	396	406
9387	MbMaleXNakakuF	m	251	251	322	330	195	215	235	235	212	238	396	402
9388	MbMaleXNakakuF	m	257	273	326	330	175	191	233	235	174	222	400	414
9389	MbMaleXNakakuF	m	257	257	322	326	175	175	233	237	218	226	396	406
9390	MbMaleXNakakuF	m	255	269	322	328	183	211	235	235	216	220	396	406
9391	MbMaleXNakakuF	m	249	257	322	330	209	215	235	243	214	226	390	394
9392	MbMaleXNakakuF	m	249	269	330	344	175	213	231	235	214	218	396	396
9393	MbMaleXNakakuF	m	241	249	318	330	183	203	223	229	218	224	396	396
9394	MbMaleXNakakuF	m	263	273	330	344	193	201	237	237	206	218	398	398
9395	MbMaleXNakakuF	m	259	259	322	322	177	177	227	237	210	224	402	414
9396	MbMaleXNakakuF	m	251	261	326	330	175	211	227	233	206	214	398	398
9397	MbMaleXNakakuF	m	271	273	322	328	175	201	235	239	174	234	396	396
9398	MbMaleXNakakuF	m	257	259	322	330	211	211	223	235	214	246	390	400
9399	MbMaleXNakakuF	m	249	253	326	332	173	213	235	237	214	224	396	408
9400	MbMaleXNakakuF	m	249	261	322	344	175	213	231	241	210	220	398	398
9401	MbMaleXNakakuF	m	253	255	328	332	175	211	235	235	218	230	396	412
9402	MbMaleXNakakuF	m	245	257	322	322	175	175	241	243	212	250	398	414
9403	MbMaleXNakakuF	m	243	249	322	332	177	201	235	235	220	220	396	400
9404	MbMaleXNakakuF	m	249	267	330	332	175	209	235	237	224	240	390	394
9405	MbMaleXNakakuF	m	257	263	322	330	173	215	235	235	210	224	396	396
9406	MbMaleXNakakuF	f	251	251	322	322	185	195	235	235	174	230	398	398
9407	MbMaleXNakakuF	f	235	269	322	330	201	201	235	237	214	216	412	416
9408	MbMaleXNakakuF	f	251	271	322	322	185	207	235	235	218	218	396	400
9409	MbMaleXNakakuF	f	251	263	316	344	199	201	235	235	200	202	392	400
9410	MbMaleXNakakuF	f	251	253	318	322	189	201	235	235	208	210	400	408
9411	MbMaleXNakakuF	f	251	257	318	334	199	203	235	235	174	244	396	400
9412	MbMaleXNakakuF	f	245	253	322	326	205	211	235	239	190	246	396	396
9413	MbMaleXNakakuF	f	245	249	322	322	201	225	233	235	204	238	396	396
9414	MbMaleXNakakuF	f	251	253	322	326	223	223	221	233	182	214	?	?
9415	MbMaleXNakakuF	f	233	243	322	328	191	225	235	235	174	214	396	400
9416	MbMaleXNakakuF	f	255	273	322	322	191	201	233	241	214	240	400	402
9417	MbMaleXNakakuF	f	253	259	322	326	193	203	221	235	222	222	390	406
9418	MbMaleXNakakuF	f	249	257	322	322	193	219	221	233	214	228	406	408
9419	MbMaleXNakakuF	f	249	271	322	322	193	193	235	235	176	234	390	396
9420	MbMaleXNakakuF	f	247	257	322	322	191	197	233	237	174	222	396	398
9421	MbMaleXNakakuF	f	247	273	322	322	195	217	221	237	214	234	396	404
9422	MbMaleXNakakuF	f	259	261	318	322	205	225	235	235	174	220	406	410
9423	MbMaleXNakakuF	f	251	253	322	322	187	205	235	235	220	224	396	416
9424	MbMaleXNakakuF	f	249	251	322	340	191	199	233	239	222	236	396	396
9425	MbMaleXNakakuF	f	251	273	318	322	203	207	229	235	190	224	392	392
9426	MbMaleXNakakuF	f	255	255	326	340	203	217	237	239	166	212	396	396
9427	MbMaleXNakakuF	f	255	257	322	332	191	201	235	235	190	214	396	406
9428	MbMaleXNakakuF	f	245	253	322	322	203	203	235	239	180	242	396	402
9429	MbMaleXNakakuF	f	249	255	322	322	195	219	235	235	174	196	396	408
9430	MbMaleXNakakuF	f	245	253	322	322	191	203	223	235	208	244	396	404
9431	MbMaleXNakakuF	f	247	259	322	322	187	203	231	235	174	228	396	396
9432	MbMaleXNakakuF	f	245	251	322	328	203	221	221	237	240	242	392	412
9433	MbMaleXNakakuF	f	235	275	322	322	195	201	239	239	222	232	396	396
9434	MbMaleXNakakuF	f	227	243	322	322	193	205	221	237	190	212	398	400
9435	MbMaleXNakakuF	f	249	255	318	322	175	201	221	235	174	222	396	396
9436	MbMaleXNakakuF	f	257	259	322	322	207	213	235	235	174	200	396	396
9437	MbMaleXNakakuF	f	251	251	322	326	175	201	229	231	170	226	396	396
9438	MbMaleXNakakuF	f	255	259	322	332	201	201	229	237	174	214	396	396
9439	MbMaleXNakakuF	f	257	265	322	322	193	201	221	235	174	202	396	396
9440	MbMaleXNakakuF	f	251	253	?	?	?	?	217	235	174	232	396	396

9441	MbMaleXNakakuF	f	243	261	318	322	207	213	235	235	210	224	396	396
9442	MbMaleXNakakuF	f	253	263	318	334	175	199	221	235	212	240	396	406
9443	MbMaleXNakakuF	f	245	257	322	322	207	211	233	239	214	222	396	398
9444	MbMaleXNakakuF	f	247	251	318	322	185	217	221	235	208	214	396	396
9445	MbMaleXNakakuF	f	255	275	322	322	207	207	221	235	174	204	396	396
9446	MbMaleXNakakuF	f	249	255	316	322	201	207	235	235	174	208	396	412
9447	MbMaleXNakakuF	f	237	263	338	340	199	207	235	235	166	174	396	400
9448	MbMaleXNakakuF	f	249	259	318	322	199	205	235	235	174	224	396	402
9449	MbMaleXNakakuF	f	255	255	318	322	203	207	235	239	174	212	392	396
9450	MbMaleXNakakuF	f	259	261	316	322	175	213	223	235	166	214	396	400
9451	MbMaleXNakakuF	f	249	259	318	328	203	207	235	235	166	174	400	404
9452	MbMaleXNakakuF	f	243	245	322	336	189	201	221	235	174	208	400	406
9453	MbMaleXNakakuF	f	251	255	324	332	175	207	221	235	220	246	396	396
9454	MbMaleXNakakuF	f	239	241	322	326	227	249	235	235	208	214	396	406
9455	MbMaleXNakakuF	f	263	267	322	322	195	207	235	235	226	236	390	390
9456	MbMaleXNakakuF	f	251	269	318	326	191	209	235	235	174	230	402	410
9457	MbMaleXNakakuF	f	255	257	318	322	201	201	233	239	174	222	396	396
9458	MbMaleXNakakuF	f	261	263	322	334	175	189	221	237	222	232	396	408
9459	MbMaleXNakakuF	f	249	251	322	334	195	203	235	235	230	246	406	410
9460	MbMaleXNakakuF	f	249	261	322	338	195	217	233	237	174	174	396	400
9461	MbMaleXNakakuF	f	233	269	322	324	195	203	221	235	174	208	396	408
9462	MbMaleXNakakuF	f	249	271	318	322	173	187	235	237	174	208	396	396
9463	MbMaleXNakakuF	f	247	247	318	322	189	193	235	239	166	220	396	402
9464	MbMaleXNakakuF	f	247	249	322	322	175	207	235	235	174	246	?	?
9465	MbMaleXNakakuF	f	245	275	322	326	189	203	219	219	166	166	390	396
9466	MbMaleXNakakuF	f	243	269	322	330	195	203	231	235	218	246	396	396
9467	MbMaleXNakakuF	f	253	261	322	338	191	201	235	239	214	234	398	406
9468	MbMaleXNakakuF	f	253	271	334	334	189	207	219	235	218	222	396	402
9469	MbMaleXNakakuF	f	245	249	318	322	195	201	219	235	174	246	390	396
9470	MbMaleXNakakuF	f	253	259	324	340	213	217	215	221	166	226	396	406
9471	MbMaleXNakakuF	f	245	259	322	324	189	197	235	235	174	190	404	406
9472	MbMaleXNakakuF	f	251	259	322	322	189	201	229	235	212	214	398	416
9473	MbMaleXNakakuF	f	253	261	316	326	201	203	231	235	174	174	396	404
9474	MbMaleXNakakuF	f	243	253	322	322	185	205	235	235	218	234	?	?
9475	MbMaleXNakakuF	f	261	263	324	340	187	207	229	233	220	224	396	396
9476	MbMaleXNakakuF	f	243	249	322	330	201	207	235	237	166	206	396	406
9477	MbMaleXNakakuF	f	241	269	316	322	189	203	221	235	174	188	396	402
9478	MbMaleXNakakuF	f	259	261	322	340	181	205	235	239	174	228	396	396
9479	MbMaleXNakakuF	f	251	253	318	322	205	211	215	235	174	224	396	406
9480	MbMaleXNakakuF	f	235	257	318	322	175	201	231	235	174	236	?	?

43

A. Table 7: Genotype table of F1 offspring

Individuum	Morph	Sex	Pzeb 2	Pzeb 2	Pzeb3	Pzeb3	UNH130	UNH130	UME002	UME002	UME003	UME003	TmoM27	TmoM27
12324	MbitaM x NakakuF	m	257	257	328	334	175	175	223	235	216	218	396	396
12325	MbitaM x NakakuF	m	231	253	324	334	177	205	235	235	216	218	398	416
12326	MbitaM x NakakuF	m	251	251	320	332	201	209	223	239	212	244	390	396
12327	MbitaM x NakakuF	m	237	269	324	324	175	201	235	237	220	220	394	410
12328	MbitaM x NakakuF	m	249	259	324	334	195	215	235	235	212	224	390	404
12329	MbitaM x NakakuF	?	247	263	324	332	197	211	235	235	188	212	398	408
12330	MbitaM x NakakuF	w	245	259	324	332	205	215	229	235	212	232	394	404
12331	MbitaM x NakakuF	m	231	255	328	330	175	225	235	243	180	232	396	396
12332	MbitaM x NakakuF	w	255	269	326	334	177	205	235	243	174	216	394	396
12333	MbitaM x NakakuF	m	243	251	324	324	201	209	235	235	188	224	?	?
12334	MbitaM x NakakuF	m	251	275	324	328	191	195	235	235	174	220	396	414
12335	MbitaM x NakakuF	m	231	237	324	332	177	201	235	243	214	234	398	412
12336	MbitaM x NakakuF	m	259	279	330	332	175	201	235	243	212	248	?	?
12337	MbitaM x NakakuF	m	231	265	324	334	177	201	235	235	200	216	396	398
12338	MbitaM x NakakuF	?	251	273	324	336	207	209	221	235	212	220	390	402
12339	MbitaM x NakakuF	m	257	259	324	336	203	211	235	237	244	246	400	400
12340	MbitaM x NakakuF	m	231	253	320	334	175	199	235	235	174	234	394	400
12341	MbitaM x NakakuF	m	255	277	324	332	175	191	235	235	206	248	398	406
12342	MbitaM x NakakuF	?	253	259	324	324	175	207	235	243	216	224	?	?
12343	MbitaM x NakakuF	m	253	255	324	324	175	211	221	235	222	248	?	?
12344	MbitaM x NakakuF	m	265	269	328	328	177	201	219	235	200	216	?	?
12345	MbitaM x NakakuF	w	257	259	324	328	175	201	221	237	174	234	?	?
12346	MbitaM x NakakuF	w	251	253	324	324	185	215	231	235	216	224	?	?
12347	MbitaM x NakakuF	w	259	259	324	324	175	201	221	235	188	248	406	414
12348	MbitaM x NakakuF	m	269	271	332	334	177	201	?	?	212	234	394	412
12349	MbitaM x NakakuF	m	259	273	324	324	195	215	221	221	174	212	390	396
12350	MbitaM x NakakuF	w	245	259	324	324	205	209	229	235	212	216	394	410
12351	MbitaM x NakakuF	m	253	259	324	324	187	215	235	235	212	218	394	396
12352	MbitaM x NakakuF	w	231	237	324	332	175	201	235	235	214	216	394	416
12353	MbitaM x NakakuF	m	231	259	330	330	175	201	235	243	174	224	?	?
12354	MbitaM x NakakuF	w	257	269	332	334	175	177	235	235	216	218	396	398
12355	MbitaM x NakakuF	m	231	237	332	332	175	201	235	243	214	216	?	?
12356	MbitaM x NakakuF	m	237	237	332	334	177	201	233	233	214	234	398	412
12357	MbitaM x NakakuF	w	251	253	328	332	207	215	235	235	174	212	390	396
12358	MbitaM x NakakuF	w	231	235	322	332	175	201	219	235	174	188	400	412
12359	MbitaM x NakakuF	w	247	253	324	332	175	195	235	235	174	208	?	?
12360	MbitaM x NakakuF	m	259	271	324	324	175	189	223	239	188	220	?	?
12361	MbitaM x NakakuF	m	231	253	324	334	177	201	235	235	216	228	396	396
12362	MbitaM x NakakuF	m	231	243	324	324	175	201	237	243	224	224	?	?
12363	MbitaM x NakakuF	m	253	269	324	324	175	175	233	235	?	?	394	394
12364	MbitaM x NakakuF	m	231	259	324	334	175	193	235	235	212	214	?	?
12365	MbitaM x NakakuF	m	253	269	324	334	175	201	235	243	224	234	394	394
12366	MbitaM x NakakuF	m	231	231	324	332	175	177	235	235	216	246	?	?
12367	MbitaM x NakakuF	w	249	259	324	332	175	191	221	235	174	246	406	414
12368	MbitaM x NakakuF	m	245	251	324	324	207	209	235	237	208	212	390	398
12369	MbitaM x NakakuF	m	231	249	324	334	175	175	221	237	234	246	398	406
12370	MbitaM x NakakuF	m	257	259	320	332	203	215	235	241	224	242	390	396
12371	MbitaM x NakakuF	w	251	259	332	340	175	195	235	235	174	220	396	414
12372	MbitaM x NakakuF	m	231	253	324	334	175	187	237	243	218	234	398	416
12373	MbitaM x NakakuF	w	255	259	332	336	189	209	235	243	212	216	394	402
12374	MbitaM x NakakuF	w	253	259	324	324	207	209	233	239	212	230	?	?
12375	MbitaM x NakakuF	m	247	277	324	324	175	203	235	239	174	242	396	404
12376	MbitaM x NakakuF	w	249	259	324	324	175	209	235	243	212	212	390	406
12377	MbitaM x NakakuF	w	259	259	324	324	191	209	229	235	174	232	?	?
12378	MbitaM x NakakuF	m	259	273	324	324	175	207	233	233	216	216	400	406
12379	MbitaM x NakakuF	w	251	253	320	332	203	209	229	233	174	212	394	400
12380	MbitaM x NakakuF	w	259	269	324	332	195	209	217	243	224	224	390	394
12381	MbitaM x NakakuF	m	259	273	324	336	189	215	235	235	216	224	390	402
12382	MbitaM x NakakuF	w	259	269	324	334	177	193	221	235	174	216	396	398
12383	MbitaM x NakakuF	w	253	269	324	332	175	221	237	241	216	240	392	398

12384	MbitaM x NakakuF	w	259	259	324	324	191	209	219	235	174	232	404	404
12385	MbitaM x NakakuF	w	263	271	324	344	201	213	221	235	174	218	?	?
12386	MbitaM x NakakuF	w	253	269	324	324	175	187	217	235	216	218	396	398
12387	MbitaM x NakakuF	w	251	257	324	332	209	217	221	235	224	232	390	404
12388	MbitaM x NakakuF	w	259	265	324	324	207	209	221	235	212	234	390	394
12389	MbitaM x NakakuF	m	251	253	324	340	175	217	235	239	174	232	400	406
12390	MbitaM x NakakuF	m	231	255	324	330	201	223	235	237	212	232	396	396
12391	MbitaM x NakakuF	w	251	271	326	332	203	209	235	235	174	224	396	396
12392	MbitaM x NakakuF	m	243	259	320	332	183	199	233	235	174	222	396	396
12393	MbitaM x NakakuF	w	259	263	324	332	175	217	235	235	174	220	400	400
12394	MbitaM x NakakuF	w	257	275	324	328	191	203	235	243	174	232	396	400
12395	MbitaM x NakakuF	m	255	259	324	328	201	215	233	237	224	230	394	396
12396	MbitaM x NakakuF	w	259	273	324	324	195	215	235	235	224	232	394	396
12397	MbitaM x NakakuF	m	263	269	332	336	175	189	235	235	214	220	396	408
12398	MbitaM x NakakuF	w	257	263	324	334	201	213	233	235	174	218	396	398
12399	MbitaM x NakakuF	w	247	251	324	324	195	215	231	235	174	224	390	396
12400	MbitaM x NakakuF	w	231	247	330	334	175	221	235	241	216	240	392	398
12401	MbitaM x NakakuF	m	251	253	324	340	201	215	235	235	212	230	390	396
12402	MbitaM x NakakuF	m	249	251	324	332	209	217	221	229	224	232	390	396
12403	MbitaM x NakakuF	w	231	259	324	334	175	201	229	235	174	234	?	?
12404	MbitaM x NakakuF	m	261	269	324	332	175	197	235	237	188	214	406	410
12405	MbitaM x NakakuF	m	231	247	330	334	175	203	235	235	216	240	392	398
12406	MbitaM x NakakuF	m	231	271	324	334	175	201	235	235	214	234	398	412
12407	MbitaM x NakakuF	m	257	259	324	328	203	209	237	239	212	212	?	?
12408	MbitaM x NakakuF	w	269	271	324	334	177	195	229	235	174	216	394	408
12409	MbitaM x NakakuF	m	251	251	324	324	209	219	235	235	194	224	394	396
12410	MbitaM x NakakuF	w	251	259	328	340	175	195	233	235	174	232	396	400
12411	MbitaM x NakakuF	w	231	253	324	328	175	225	235	235	174	180	?	?
12412	MbitaM x NakakuF	w	253	259	324	328	191	209	235	235	174	212	390	410
12413	MbitaM x NakakuF	w	239	243	328	342	187	199	223	235	174	216	396	406
12414	MbitaM x NakakuF	w	253	269	328	332	175	177	235	243	168	234	394	396
12415	MbitaM x NakakuF	w	245	247	324	330	175	191	235	235	174	248	400	414
12416	MbitaM x NakakuF	m	231	259	324	334	175	193	239	239	200	216	396	398
12417	MbitaM x NakakuF	w	243	253	324	332	183	223	235	243	212	222	396	398
12418	MbitaM x NakakuF	m	257	275	324	332	175	203	235	241	212	220	396	414
12419	MbitaM x NakakuF	m	231	271	324	332	177	201	235	237	212	234	398	412
12420	MbitaM x NakakuF	m	251	263	320	324	209	213	233	237	222	224	390	398
12421	MbitaM x NakakuF	m	249	259	324	332	175	207	235	237	174	246	406	414
12422	MbitaM x NakakuF	w	257	269	328	332	175	203	233	237	166	220	396	396
12423	MbitaM x NakakuF	m	253	269	324	324	177	201	235	235	170	216	396	396
12424	MbitaM x NakakuF	w	251	255	332	340	201	215	235	235	174	212	394	396
12425	MbitaM x NakakuF	m	251	259	324	332	191	217	235	237	174	234	396	400
12426	MbitaM x NakakuF	w	231	251	324	334	175	195	235	239	174	234	398	408
12427	MbitaM x NakakuF	w	251	251	324	332	195	209	219	235	174	212	394	396
12428	MbitaM x NakakuF	?	253	261	320	332	167	199	235	235	174	212	390	400
12429	MbitaM x NakakuF	w	251	255	324	328	201	215	221	243	174	224	394	396
12430	MbitaM x NakakuF	w	251	251	324	324	195	209	235	235	224	232	390	398
12431	MbitaM x NakakuF	?	271	279	324	324	175	201	233	235	220	220	?	?
12432	MbitaM x NakakuF	m	251	253	332	336	203	209	219	235	174	224	390	400
12433	MbitaM x NakakuF	w	257	259	320	324	175	203	235	237	210	244	400	414
12434	MbitaM x NakakuF	m	231	273	324	334	175	201	237	239	212	234	398	416

12435	MbitaM x NakakuF	m	257	259	324	328	191	203	221	235	174	212	396	414
12436	MbitaM x NakakuF	w	253	269	324	334	199	207	235	235	218	228	?	?
12437	MbitaM x NakakuF	m	257	259	324	328	203	209	235	243	212	212	390	396
12438	MbitaM x NakakuF	m	231	241	322	334	177	227	235	243	206	216	400	406
12439	MbitaM x NakakuF	?	253	259	324	332	195	215	235	235	174	224	390	396
12440	MbitaM x NakakuF	?	261	277	324	324	195	209	233	235	212	224	390	396
12441	MbitaM x NakakuF	?	261	279	324	324	175	191	235	235	212	248	396	406
12442	MbitaM x NakakuF	?	249	261	320	332	195	215	233	235	224	244	390	394
12443	MbitaM x NakakuF	?	235	255	324	324	175	195	235	235	206	206	396	396
12444	MbitaM x NakakuF	?	249	259	324	328	209	217	235	239	224	232	404	404
12445	MbitaM x NakakuF	?	251	273	318	324	201	209	235	235	188	224	390	390
12446	MbitaM x NakakuF	?	243	265	332	342	187	199	235	235	174	216	396	398
12447	MbitaM x NakakuF	?	?	?	324	342	175	217	235	235	174	224	?	?
12448	MbitaM x NakakuF	?	257	269	324	342	175	217	235	235	174	210	396	410
12449	MbitaM x NakakuF	?	257	269	332	342	175	203	235	239	210	220	396	410
12450	MbitaM x NakakuF	?	253	277	324	328	175	209	235	239	174	248	396	402
12451	MbitaM x NakakuF	?	251	251	324	332	195	215	235	239	174	224	394	396
12452	MbitaM x NakakuF	?	259	271	324	328	209	209	237	241	174	212	390	410
12453	MbitaM x NakakuF	?	251	273	324	328	195	215	235	235	212	232	396	396
12454	MbitaM x NakakuF	?	251	273	324	332	195	209	235	235	174	224	394	396
12455	MbitaM x NakakuF	?	249	259	324	332	191	207	235	239	174	248	?	?
12456	MbitaM x NakakuF	?	253	261	324	334	201	209	235	239	244	234	?	?

Printed by Books on Demand GmbH, Norderstedt / Germany